Kefas Babale Shitta

Babesia Canis Infection and Its Vector

AF153188

Kefas Babale Shitta

Babesia Canis Infection and Its Vector

Canine babesiosis

LAP LAMBERT Academic Publishing

Impressum / Imprint

Bibliografische Information der Deutschen Nationalbibliothek: Die Deutsche Nationalbibliothek verzeichnet diese Publikation in der Deutschen Nationalbibliografie; detaillierte bibliografische Daten sind im Internet über http://dnb.d-nb.de abrufbar. Alle in diesem Buch genannten Marken und Produktnamen unterliegen warenzeichen-, marken- oder patentrechtlichem Schutz bzw. sind Warenzeichen oder eingetragene Warenzeichen der jeweiligen Inhaber. Die Wiedergabe von Marken, Produktnamen, Gebrauchsnamen, Handelsnamen, Warenbezeichnungen u.s.w. in diesem Werk berechtigt auch ohne besondere Kennzeichnung nicht zu der Annahme, dass solche Namen im Sinne der Warenzeichen- und Markenschutzgesetzgebung als frei zu betrachten wären und daher von jedermann benutzt werden dürften.

Bibliographic information published by the Deutsche Nationalbibliothek: The Deutsche Nationalbibliothek lists this publication in the Deutsche Nationalbibliografie; detailed bibliographic data are available in the Internet at http://dnb.d-nb.de. Any brand names and product names mentioned in this book are subject to trademark, brand or patent protection and are trademarks or registered trademarks of their respective holders. The use of brand names, product names, common names, trade names, product descriptions etc. even without a particular marking in this work is in no way to be construed to mean that such names may be regarded as unrestricted in respect of trademark and brand protection legislation and could thus be used by anyone.

Coverbild / Cover image: www.ingimage.com

Verlag / Publisher:
LAP LAMBERT Academic Publishing
ist ein Imprint der / is a trademark of
OmniScriptum GmbH & Co. KG
Heinrich-Böcking-Str. 6-8, 66121 Saarbrücken, Deutschland / Germany
Email: info@lap-publishing.com

Herstellung: siehe letzte Seite /
Printed at: see last page
ISBN: 978-3-659-66448-9

Zugl. / Approved by: Jos, University of Jos, Disseertation, 2009

Acknowledgement

Every accomplishment in life is a result of the contribution of many individuals who both directly and indirectly share their gifts, talents, and wisdom with us all. This project is no exception. I am indepted to my dear wife, Jemimah, my parents, Jibrin Babale Shitta and Abisha Babale Shitta, my siblings; Manasseh, Effor, Bwecha and Tibwa for their limitless support and encouragement. I also want to thank my mentors; Professor Oladele Benjamin Akogun and Professor Naomi Naret James-Rugu who both mentored and taught me on how to write a report on a research carried out and present same in a scientific conference. I appreciate my publishers greatly, particularly Olivia Morrison who saw one of my work online and showed interest in publishing the work for a wider world.

Table of Content **Page**

CHAPTER ONE

INTRODUCTION AND LITERATURE REVIEW

1.1 Introduction

The dog is one of the most popular pets in the world. It ordinarily remains loyal to a considerate master, and because of this the dog has been called man's best friend. Class distinctions between people have no part in a dog's life. It can be a faithful companion to either rich or poor. Dog has been man's oldest friend and shares a lot of thing in common including happiness grieve and diseases. Many families in most parts of the world have one or more dogs in their homes. Dogs naturally protect their homes and will resist intrusion by either barking or outright attack especially when they consider their territory being threatened (Obori, 2001 and Mery, 2006).

Dogs are also very useful in sports and hunting. The great potentials inherent in dog breeding, especially in income generation and employment can never be overemphasized. Many families have realized substantial income from breeding dogs to support family upliftment (Obori, 2001). Dogs have been domesticated for most of human history and have thus endeared themselves to many over the years. Stories have been told about brave dogs that served admirably in war or that risked their lives to save persons in danger. When Pompeii--the Roman community destroyed by Mount Vesuvius in AD 79--was finally excavated, searchers found evidence of a dog lying across a child, apparently trying to protect the youngster. Perhaps few of the millions of dogs in the world may be so heroic, but they are still a source of genuine delight to their owners (Mery, 2006).

1.2 Literature Review

Dogs exist in a wide range of sizes, colors, and temperaments. Fig.1. some, such as the Doberman pinscher and the German shepherd, serve as alert and aggressive watchdogs. Others, such as the beagle and the cocker spaniel, are playful family pets, even though they were bred for hunting. Still others, such as the collie and the Welsh corgi, can herd farm or range animals. Each of the dogs just mentioned is a purebred. A mongrel dog, however--one with many breeds in its background--can just as easily fit into family life. Dogs are trained as guard dogs in peacetime by the United States Army and other military services. Because of their keen sense of smell, dogs are used by police at times to track down escaped prisoners. Law enforcement agencies also rely on the dog's acute sense of smell to uncover illegal drugs. And specially trained dogs serve as the "eyes" of the blind, guiding the steps of their sightless masters around obstacles and hazards. Doctors used dogs to determine whether a person was dead or in a coma, a wag of the dog's tail would indicate life, but a silent dog meant the person was indeed dead (Mery, 2006)

Fig. 1. Two Alsatian dogs at the study site

Ticks have been reported to infest a wide range of mammalian hosts hence they are referred to as eurixenous ectoparasites (Harwood and James, 1969). It is a well-known fact that dogs at one stage or the other are infested with Ixodid ticks (Hoogstraal, 1956) ranging from *Rhipicephalus, Boophilus*, and *Haemaphysalis* species.

1.2.1 Ticks

Ticks are ectoparasitic arthropods. All ticks are parasitic on terrestrial vertebrates (mammals, birds, and reptiles) during at least one stage in their life cycle and are also known as dreaded blood sucking, external parasites that pose greater risk on both man and dog (Farley, 1996). Approximately 850 species have been described worldwide. There are two well established families of ticks, the Ixodidae (hard ticks), and Argasidae (soft ticks). Both are important vectors of disease causing agents to humans and animals throughout the world. Ticks transmit the widest variety of pathogens of any blood sucking arthropod, including bacteria, rickettsiae, protozoa, and viruses. Some human diseases of current interest in the United States caused by tick-borne pathogens include Lyme disease, Ehrlichiosis, babesiosis, Rocky Mountain Spotted Fever (RMSF), tularemia, and tick-borne relapsing fever (Vredevoe, 2006).

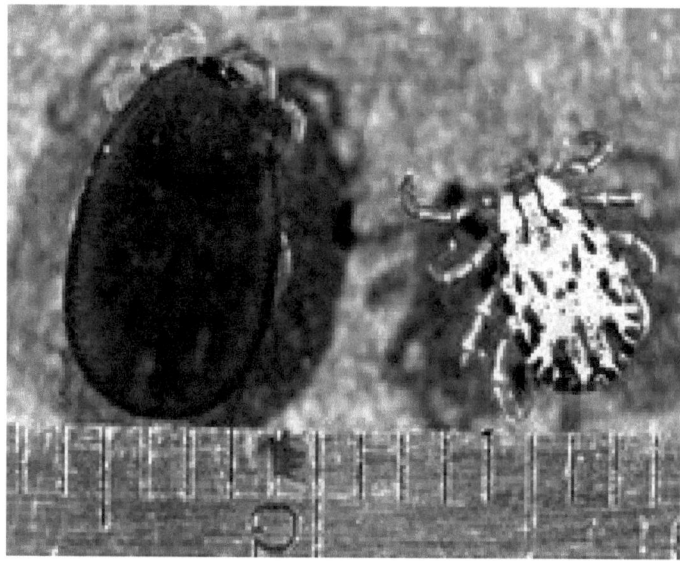

Fig. 2. Soft tick (left) and Hard tick (right)

Soulsby (1973) noted that ticks may harm their host through; injuries done by their bites, sucking blood and transmission of viruses, rickettsial, bacterial and protozoal infections.

1.2.2 Classification of ticks
Ticks belong to the phylum Arthropoda.

Characteristics of phylum Arthropoda. Soulsby (1982) reviewed the diagnostic features of arthropods as follows:
1. Possession of exoskeleton.
2. Possession of bilaterally symmetrical body.
3. Presence of paired jointed appendages.
4. Possession of epidermal cuticle consisting of chitin and protein.
5. Possession of compound eyes in some major groups.
6. Possession of open circulatory system.
7. Presence of segmented body.
8. Possession of nervous system, with an anterior ganglion and brain.

Characteristics of Class Arachnida

The features of this class include:
1. Antennae, wings and compound eyes absent.
2. Their mouth parts are small; they suck the tissue of their host by means of pharynx.

3. Segmentation of arachnids differs from other arthropods and described as prosoma and opistosoma for first segment and remaining segment respectively.
4. Most of them possess poisonous glands.
5. Arachnids breathe by means of the gill-books, lung-book and trachea (Soulsby, 1982).

Characteristics of Order Acarina

The features of this order include:
1. Mouth parts made up of chelicerae and pedipalps.
2. Mouth parts attached to false head known as the gnathosoma.
3. Antenae is absent.
4. The body is sac-like without any segmentation.
5. Adult and nymphal stages possess four pairs of legs while the larval stage that hatched from the egg has only three pairs of legs.

Characteristics of Sub-order Ixodoidea

Two families have been identified in the sub-order Ixodoidea, the Argasidae and Ixodidae (Nelson 1975).

Argasidae (soft ticks) have the following characteristics:
1. Absence of scutum.
2. Their mouth parts originate on ventral surface and are not visible on the dorsal surface as is the case for members of the Ixodidae family.

Ixodidae (hard ticks) have the following characteristics:
1. Possession of visible scutum.
2. Mouth parts originate on the anterior margin and are visible on the dorsal surface.

Table 1: Broad Classification of ticks

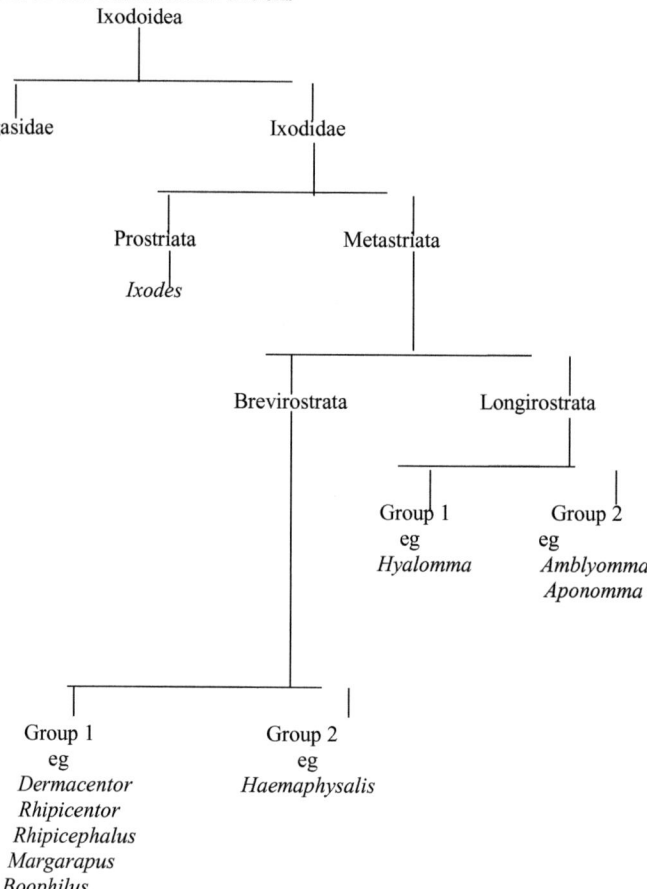

1.2.3 Life cycle of Ixodid tick

The female lays about 3,000 eggs. The egg is spherical in shape and is visible to the naked eye. Hatching occurs after a period varying from a few weeks to several months and the newly emerged tick ("seed tick") is only about 0.5 to 1.5mm long and has three pairs of legs.

It attaches to a suitable host after a few quiescent days after a blood meal, drops to the ground to molt into a nymph with four pairs of legs. The nymph seeks a new host, and engorges itself with blood which it digests in a sheltered corner. It then matures to an adult after a final molt (Vredevoe, 2006 and Foster and Smith, 2008).

It was noted by Lancaster and Meisch (1986) that the female engorges to repletion, mating takes place on the host and the female drops to the ground. When replete she may be as small as a bean or as large as a grape, and vary in color from yellowish brown to blue grey. She seeks shelter either on the surface of the soil in a crack or crevices or beneath leaves in tuft of grass. Within a week to 10 days depending on the environmental conditions, she begins laying a mass of several thousands of eggs. When finished the female then dies. The eggs developed within several weeks or even mouths depending also on environmental conditions (and the species involved), it then hatched into 6-legged larvae or seed tick.

Lancaster and Meisch (1986) in their studies suggested that these 'seed ticks' soon climb a blade of grass, forming cluster and wait for a passing host. This host seeking behavior in ticks is called "questing" The larvae climbs into suitable host and attaches, and feed to repletion in a few days. In some species, the larvae remain on the host and then molt to the nymph. In other species the larvae drops to the ground.

Fig.3 Questing   in   HardTick
       (Ixodes scapularis)

The nymph, if already on the same host, attaches, feeds; either molts to the adult or may drop of in some species to molt on the ground. In some other species the nymph, has to find a new host on which to feed before completing its engorgement and dropping of an adult on the ground. The emerging adult from a nymph on the same host simply reattaches, mates, feeds to repletion and drops to the ground to start the cycle over again. In other cases, the adult has to seek a new host for feeding, mating and engorgement preparatory to dropping on the ground and starting the cycle again. These three different patterns of feeding, molting and being on the host or off the host led to a classification of species as 1- host, 2-host, or 3-host ticks. The 1- host ticks spend their entire development period on host animal from the time they attach larvae until the replete female drops off the ground. The 2-host ticks attach as larvae, complete their development through the engorged nymphs which drop off to molt. The adult has to then find different host on which to complete its development. The 3-host ticks' attaches as larvae engorge drop to the ground and molt to the adults which have to find still another host for feeding before being capable of producing the viable eggs for the new generation (Lancaster and Meisch, 1986 and Vredevoe, 2006).

According to Hall (1994) the eggs are laid off the host 2,000 to 20,000 depending on the environmental conditions and species. They will hatch quickly and all at once under conditions of high temperature and moderate humidity, but dry conditions desiccate them. Larvae climb to the top of grand blades or other vegetation, and attaches to passing host. From then the life cycle differs according to whether it is a 1-host ticks as in the case of *Boophilus spp*, 2 – host ticks as in *Rhipicephalus, Haemaphysalis* and *Hyalomma spp* or 3 – host ticks as in *Amblyomma* and some species of *Rhipicephalus, Haemaphysalis* and *Hyalomma*.

Hall (1994) suggested that in the case of the 2 and 3 host-ticks the individuals selected as hosts may be of the same or different species, host specificity not being a characteristic of ticks. Diseases are transmitted through the 1 – host tick only transovarially. A 2- host tick feeding on the same species of host can transmit a disease directly. A 3 – host tick, feeding on the same species can also transmit a disease directly.

1.2.4 Life cycle of Argasid tick

Argasid ticks are ticks of warm countries. Hoogstraal (1956) gives the detail life cycle of Argasid ticks under laboratory conditions ($27^{\circ}C - 32^{\circ}C$ and $40 - 50\%$ R.H.) as: The eggs being laid in batches of 30-50 and hatch in 16-20 days. The larvae usually feed for 17-19 days and 5-10 days later molt to become nymphs. They are capable of feeding within three to four days of molting and they become replete within 20 to 30 minutes. The nymphs usually have two blood meals. Each of these is followed by a molt, even though the males may emerge after the first molt. Males and females may feed within seven days of molting and the duration of each adult feed is 30-40 minutes.

In the first six months of adult life eggs are not laid, however, during this period both sexes may consume several meals of blood.

Nnochiri (1975) reports that the life-cycle of ticks is hemimetabolous, meaning, the newly hatched larvae closely resembles the parents and this resemblance continues throughout the various molts; the difference between the adult and immature stages therefore, lie in their size and sexual maturity. This is also reported by Foster and Smith (2008).

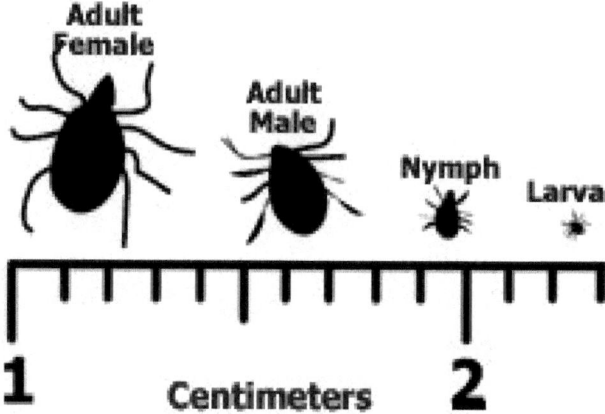

Fig. 4. Nature of resemblance in life cycle of a tick.

Nnochiri (1975) further noted that eggs which are laid in cracks and crevices of walls and floors after fertilization following an essential blood meal may number over 1000. Egg hatch in 1 to 4 weeks and the hexapod larvae that emerges molts to pass through four nymphal stages before maturing into an adult. Depending on the availability of food and the species, a life cycle last 6 to 12 months.

Hall (1994) noted that the life-cycle of soft tick differs from that of the hard ticks in several features. The larvae are not parasitic on animals in some species while, in others they are parasitic. In some species there are a number of molts of the nymphs, a blood meal is usually taken on the same or different individuals between each molts. Mating takes place off the host and the female feeds drops off and lays eggs, repeating the process a number of times. Feeding times are measured in hours or even minutes rather than days, and the time between feeds, which may be weeks or even months, is spent in molting or laying eggs in the habitat of the host which is usually well hidden. One exception to this general life cycle is the spinose ear tick (*Otobius megnini*) in which the larvae and nymphs remain in the host's ear for some months; engorged nymphs then drop to the ground, molt, mate and the female lays eggs without further blood meal.

1.2.4 Morphology of ticks

The integument consists of an outer cuticle and a single layer of epithelial cells that secrete it as is the case in other arthropods. The cuticle may be membranous or leathery, sometimes has hard plates or shields, special structures such as glands, setae and sensory organs are derived from integumentary cells (Daburum, 2005). Robert and Janovy (1996) observed that tagmatisation has resulted in two body regions; an anterior gnathasoma, or capitulum bearing the mouth parts and a single Idiosoma, containing most internal organs and bearing the legs. The Idiosoma is further divided into regions as thus; the portion bearing the legs is podosoma, the first and second pairs of legs are on the prodosoma and the third and fourth pairs are on the metapodosoma. The portion of the body posterior to the legs is the opisthosoma. The gnathasoma and prodosoma together comprise the proterosoma, and the metapodosoma and opisthosoma together are the hysterosoma. These terms are useful in describing and the identification of acarines (Roberts and Janony, 1996) also reported by (Daburum, 2005).

They further stated that the capitulum mainly is made up of feeding appendages surrounding the mouth, on each side of the mouth is a chelicerae, which functions in piercing, tearing, or gripping host tissues, the form of the chelicerae varies greatly in different families; thus they are useful taxonomic features. Lateral to the chelicerae is a pair of segmental pedipalps, which also vary greatly in form and function related to feeding, ventrally the coxae of the pedipalps are fused to form a hypostome, a rostrum, or rectum extends dorsally over the mouth. Some or all of these structures can be retracted in some ticks (Roberts and Janovy, 1996) also repoted by (Daburum, 2005). The mouthparts of hard ticks are readily visible from above. There are three visible components: the two outside jointed parts are the highly mobile **palps**; between these are paired **chelicerae**, which protect the center rod-shaped structure, the **hypostome**. The palps move laterally while the tick is feeding and do not enter the skin of the host. The rough hypostome has many beak-like projections on it. This is the structure which plunges into the host's skin while feeding. The backward directed projections prevent easy removal of the attached tick. In addition, most hard ticks secrete a cement-like substance produced by the salivary glands which literally glues the feeding tick in place; the substance dissolves after feeding is complete.

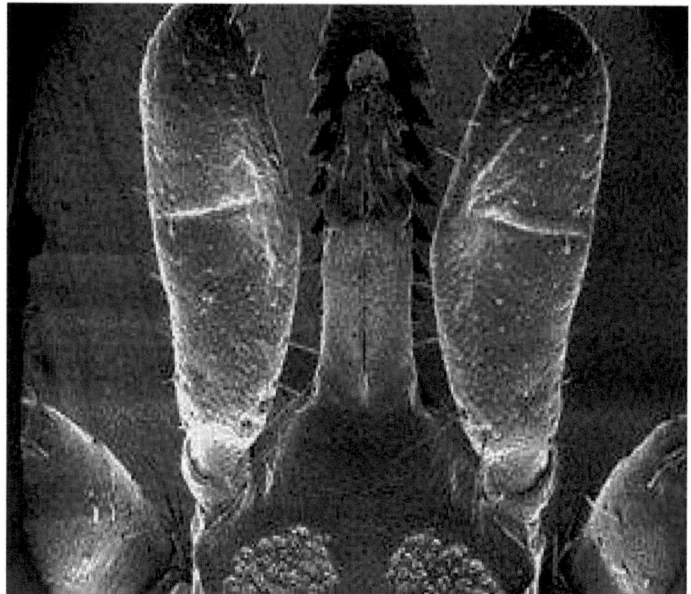

Fig. 5. Scanning electron micrograph of tick mouthparts

1.2.5 Medical importance of ticks

Ticks are primarily parasites of wild animals and only about 10% of the species feed on domestic animals. But, these relatively few species have prospered on domestic livestock and are responsible for considerable loss to farmers in developed countries as well as to rural herdsmen in developing countries (Zeleke and Bekele, 2004).

Ticks are known to transmit the widest variety of pathogens of any blood sucking arthropods, including bacteria, rickettsiae, protozoa and viruses. Some human diseases of current interest in the United States caused by tick-borne pathogens include Lyme disease, Ehrlichiosis, 'tick fever' (Babesiosis and Anaplasmosis) (Hall, 1994 and Foster and Smith, 2008).

Tanwia (1989) indicated that the female dogs carried greater loads of ticks than the male dogs, dogs with dark fur condition were preferred by ticks and therefore have higher infestation (Ombugadu, 1993). In Africa, as in other tropical and sub-tropical regions of the world, ticks and tick-borne disease have, in addition to other socio-economic parameters, constituted major impediments to the development of an economically viable livestock industry (Mohammed and Agbede, 1980; George, 1987; De-Castro, 1997 and George 2003). As vectors of important haemoparasitic disease, ticks surpass all other arthropods, apart from mosquitoes, as disease agents (Hoogstraal, 1978).

1.3.1 *Babesia canis*

Babesia canis is a hematozoan parasite that is transmitted by ticks to wild and domestic dogs. The disease condition is called Babesiosis of the genus *Babesia*. It ranges from a relatively mild to a fatal disease. It has a relatively large size (2.4 $^\mu$m x 5.0 $^\mu$m), piriform shape and can appear in red blood cells singly or in groups of 2.4 or even more (Camacho *et al.*, 2001).

Two out of the 73 identified *Babesia species* are known to cause natural infection in dogs; *Babesia canis* and *Babesia gibsoni*.

Of the two, *Babesia canis* is considered to be the most important species affecting dogs and it occurs worldwide. It is common in tropical and sub-tropical parts of the world including India. It can be sub-clinical, acute, or chronic. Clinical signs consistent with canine babesiosis include depression, weakness, anorexia, mucous membranes, icterus, pyreseia and splenomegaly. Although haemolytic anemia is the hallmark of the disease, a number of variations like thrombecytopenia, bilirubinemia, bilirubinuria, hemoglobinuria, cellular and granular casts in the urine sediment and azotemia might occur. Disseminated intravascular coagulation and metabolic acidosis may develop as complications (Edward *et al.*, 1983).

1.3.2 Life cycle of *Babesia canis*

When a tick ingests the parasites most of them die in its intestine but some become vermiform and enter the intestinal cells, there they grow into large amoeboid form, multiply by a series of binary fission and produce over 1000 individuals in 2-3 days. These become vermiform and pass into the body carefully. They do not develop in the larval tick that hatches from egg but when it molts to the nymphal stage the parasites enter the salivary gland, multiply by a series of binary fissions; they fill each host cell with thousands of minute parasites. These become vermiform; breakout of the host cell enters the lumen of the salivary gland and is injected into the vertebrate host by the nymph or adult tick.

In the vertebrate the parasites enter the erythrocytes of the vertebrate host divide by binary fission and eventually destroy the affected red blood cells. There are no sexual stages (Cleveland, 2003 and Smyth, 2005).

1.3.3 Epidemiology of babesiosis

Babesiosis is common to domestic and wild dogs. In some cases transmitted to man and other animals by many species of the hard bodied ticks of the family ixodidae. The tick may act as vector and reservoir of infection. Hall (1994) observed that the vector ticks may be 1- host, 2 – host or 3 – host. The parasites must divide several times within the ticks and enter salivary glands before the tick becomes infective (White, 1998).

1.3.4 Clinical manifestation of babesiosis

The disease can be clinically classified into uncomplicated and complicated forms. Uncomplicated cases typically present with signs relating to acute haemolysis, including fever, anorexia, depression pale mucous membranes, splenomegally, and a water hammer pulse. This form is further derided into mild, moderate or severe disease according to the severity of the anaemia (Edward *et al.*, 1983 and Camacho *et al.*, 2001).

The complicated form of babesiosis refers to clinical manifestations that are not easily explained by the haemolytic disease process alone possible complications included acute renal failure, cerebral signs, coagulopathy icterus and hepathopathy, immune mediated haemolytic anaemia, acute respiratory distren syndrome, haemoconcentration, hypotention cardiac involvement, and pancreatitis. Overlap between the different categories of the complications can also occur (Farwell, 1982).

1.3.5 Occurrence of babesiosis in man

Since the time that the first case of human babesiosis was documented in Yugoslavia (Skrabalo *et al.*, 1957) numerous cases of human babesial infections have been reported in Europe and United States (Shih *et al.*, 1997). The symptoms of human babesiosis consist primarily of prolonged fever and recurrent chills. There may be cephalgia, myalgia, abdominal pains, anaemia, jaundice, haemoglobinuria. Most of these infections can lead to death (Hunfeld *et al.*, 2002).

1.3.6 Occurrence of babesiosis in animals

The symptom of babesiosis in different species of animals is similar, it can have a benign course with spontenouse recovery, or it may progress into a debilitating disease often ending in death (Farwell *et al.*, 1982).

The disease is manifested by fever, anorexia, followed by weakness and staring coat. In hyperacute cases, death may occur at this stage. There is lacrymation and diarrhea. Anaemia may also result which is as a result of erythrocytic destruction. At about the third day, hemoglobinuria is observed, the urine becomes dark-brown in colour. Increase heart rate and respiration become evident and are followed by pallor, becoming extremes as the red cells are progressively destroyed. The heart beat may be so forceful that it may be detected some distance from the animal. In protracted or less severe cases, jaundice and constipation follow. Yellowish faecal matter and emaciation then follow. Abortion occurs in pregnant animal (Edward *et al.*, 1983).

According to Edward *et al.*, (1983) autopsy reveals an enlarged spleen with a dark red splenic pulp of a soft consistency, an enlarged yellowish-brown liver, slightly oedematous lungs and haemorrhagic serous fluid in the pericardial cavity. Parasitemia may persist for several months to years in the absence of clinical symptoms. This latent infection can be reactivated by stress such as intercurrent infections splenectomy, or immunosuppressive therapy.

Immunity to *Babesia canis* in animals is thought to be related to persistent parasitemia. Thus, most animals are resistant to re-infection until the initial infection has been cured spontaneously or by appropriate therapy. Not much is known about immunity to Babesial infections in humans.

Canine babesiosis is one of the commonest protistan diseases of dogs which are of zoonotic importance in tropical and sub-tropical parts of the world including India (Bansal *et al.,* 1985) and the tropical Nigeria (Oduoye and Dipeolu, 1976). *Babesia canis* was first recorded by christophers (1904) and later by James (1905a) as an important disease of dogs, causing heavy mortality in imported breeds (Shortt, 1936). Banersee (1933) also reported four cases of *Babesia canis* infection in dogs. From the continental United States canine babesiosis was first reported in 1934 Losos (1986) as the first pathogenic protozoan shown to be transmitted by an arthropod vector but, ticks are the main vectors. Fever due to *Babesia canis* was reported by Duckworth (1944).

Babesia canis has been reported to have a worldwide distribution and affect a significant percentage of dog population (Farwell *et al.,* 1982; Edward *et al.,* 1983; Camacho *et al.,* 2001; Baneth *et al.,* 2003 and Oyamada *et al.,* 2005). It is one of the tick-borne blood parasites of dogs commonly encountered in the Northern States of Nigeria (Oduoye and Dipeolu 1976; Leeflang *et al.,* 1977; Odewumi *et al.,* 1979 and Eseyin 1996).

Babesia canis is a large (2-4 x 4-7 µm) protozoan usually seen as paired piriform trophozoites in affected red blood cell (Farwell *et al.,* 1982 and Bansal *et al.,* 1985). They are highly host specific. *Babesia canis* may coexist with *Ehrlichia canis* in some dogs where concurrent infections cause a severe often fatal disease (Edward *et al.,* 1983).

Two out of the 73 identified *Babesia species* are known to cause natural infection in dogs; *Babesia canis* and *Babesia gibsoni*. Of the two *Babesia canis* is considered to be the most important species affecting dogs with wider geographic distribution than *Babesia gibsoni* (Losos, 1986 and Camacho *et al.,* 2001).

Rudzinska (1981) observed that merozoites and trophozoites are the two developmental forms of *Babesia sp* which occur in the erythrocyte. *Babesia* species seem to feed by means of special organelle which is composed of tightly coiled double membranes located partly inside and partly outside the parasite.

The feeding mechanism in extra cellular *Babesia* was investigated and it was suggested that haemoglobin is pinocytosed (Langreth, 1976), but unlike malaria, *Babesia* does not have haemoglobin feeding vacuoles of haemozoin. The ultra-structural differences, between the intra-erythrocytic stages of *Babesia gibsoni* and *Babesia canis* are not qualitative but quantitative; i.e. some organelles especially the endoplasmic reticulum, are better developed in *Babesia canis* (Losos, 1986).

Dogs infected with *Babesia canis* develop accumulations of parasitized red blood cells in the brain, eyes and periorbital tissues as well as in the spleen, kidneys, skeletal muscles, intestine and lymph nodes. Haemoglobinuria was not observed, but, histologically the disease was characterized by haemosiderin deposits in the liver, spleen, bone marrow and other reticulo-endothelial cells (Smyth, 2005).

Losos (1986) reported that *Babesia* species is variable morphologically but the characteristic intra erythrocytic form is paired with piriform bodies joined at their pointed ends to form an acute angle. These piriform bodies vary considerably in size among species.

1.4 Aim and Objectives of the study

Dogs are kept as pets, apart from this some dog owners keep these carnivorous animals for the purpose of security thus, increasing the interaction between humans and dog. Considering the increasing interaction between humans and dogs, man stands the risk of being infected with haemoparasites as reported by Hoogstraal (1956) Harwood and James (1969) that ticks are potential vectors of a wide range of disease agents which range from protozoans, rickettsial, bacterial and spirochaetal infections.

This work is entirely on protozoan infection with particular reference to *Babesia canis*. Hence the objectives are as follows:

1. To determine the species of ticks that commonly infests dogs.

2. To relate the infestation of ticks with host factors such as mode of life of the host.

3. To determine the monthly abundance of tick infestation.

4. To determine the Packed Cell Volume (PCV) of the infected dogs and compare the values with those of non-infected dogs.

5. To ascertain the infection rate of *Babesia canis* and relate the infection to host factors such as Age (young/Adult), Sex, breed and mode of life.

CHAPTER TWO
MATERIALS AND METHODS

2.1 Study area

This study was carried out in Jos North and Jos South Local Government areas of Plateau State. Plateau State is almost in the center of Nigeria it lies between latitude 7° and 25° E, Plateau occupies an area of about 53,585sq kilometers (Ministry of Information Plateau State 1986) also reported by James-Rugu (2001). It also lies at an altitude of 1,280m above sea level with many granite hills. It is located in the Northern ginuea savannah vegetation belt (Keay, 1959) and has an annual rainfall of about 152cm-205cm. Concentrated almost entirely to a five months period from May to September. Maximum temperature ranges from about 24°C in August to 32°C in April, while the minimum temperature range between 14°C in December and 25°C in April.

Samples for this research work were collected from two veterinary clinics, one at Jos North and one at Jos South. The two towns were arbitrarily chosen to represent a fair sample of range and Babesial infections and its vectors in dogs in parts of Plateau State. The Jos North veterinary clinic is Government owned which is situated at the western part of Jos and is about 5km away from the Bauchi road campus of the University of Jos. It is found adjacent the polo club ground very close to the 'Y' Junction before "Gada biyu" ward when going to "Farin gada" ward all in Jos. The Jos South veterinary clinic is a private clinic owned by the Evangelical Church of West Africa (ECWA) is about 13km away from the Bauchi road campus of the University of Jos, and is located in Bukuru opposite the Junction leading to Theological College of Northern Nigeria (TCNN) in Jos South.

2.2 Physical examination of dogs

It is important to examine the dogs for the presence of ectoparasite, the dogs were therefore examined for ectoparasite before collection of blood.

2.3 Sampling method

Three hundred dogs were randomly selected from those visiting the Plateau State Veterinary Clinic and ECWA Veterinary Clinic for routine checkup. Samples were collected from unclinically sick dogs mixed and Alsatian. The dogs were divided into three age-groups namely: Young (less than 5 months old); Adolescent (5 months to 2 years old exclusive); and Adult (2 years and above). The sexes of the dogs were noted; and so also the mode of life of the dog. (Appendix IV)

2.4 Collection of blood sample

A piece of cotton wool was soaked in 70% alcohol and used to clean the site of the saphenous vein on the front limb of the dog after the dog's mouth had been covered and tied with mouth guard. A sterile 5ml syringe and hypodermic needle was pierced into the saphenous vein and about 3ml of blood drawn out from the vein of each dog. During the course of blood taking, the upper part of the vein was held tight so as to enhance the forceful flow of blood into the syringe. The blood samples were subsequently placed in labeled cleaned specimen bottles containing the anticoagulant Ethylene Diamine Tetra-acetic Acid (EDTA) and taken to the laboratory after the blood has been allowed to mix properly with the anticoagulant.

2.5 Preparation of blood smears

During the parasitological screening in the laboratory, thin and thick blood smears were made from the blood samples. Thick blood smears were prepared by putting a drop of blood on a clean grease-free microscopic slide, and the blood was spread into a circular layer with the aid of a glass rod and allowed to air dry, in a flat, level position protected from flies, dust and extreme heat (WHO, 1991). Code-number was then written on the glass slide which corresponded with the code number of the dog being examined.

Thin blood smears were prepared by putting a drop of blood on a clean grease-free microscopic slide at a distance of about 1cm from the edge of its length. After placing the slide on a flat, form surface, another clean slide with a smooth edge was used as a "spreader" to spread the blood. The smooth edge of the slide was placed just in front of the drop of blood; the blood was allowed to run along its edge. As the blood spread along the edge of the slide (spreader) it was then pushed forward rapidly and firmly at an angle of 45°, so that a thin blood film with a distinct head, body and tail region was left behind (WHO, 1991). The code number was written on it and allowed to air dry in a flat, level position protected from flies, dust and extreme heat.

2.6 Fixation of blood smears

The blood smears were fixed in ethanol for about 5 minutes and then allowed to dry at room temperature.

2.7 Staining of blood smears

The fixed blood smears were then stained with Giemsa stain for about 30 minutes inside a staining box. They were then washed properly with distilled water. The stained smears were left to dry at room temperature in a slide-rack placed in a standing position.

2.8 Examination and identification of blood parasites

The thick blood smears were examined first and if any of the slides was positive, the corresponding thin film was used to identify the parasites properly.

The slides were examined under a research microscope fitted with an oil-immersion objective lens (x100). The time of examination of each blood sample on a slide range from 10-20 minutes. If parasites were observed, more time was spent on the examination of the slide in order to identify the parasite density properly.

For babesial parasites, the main body of the smear was observed, as the red blood cells were evenly spread in this area. An organism with red chromatin, bluish cytoplasm found in the red blood cells confirmed the parasite's presence. The parasite could be amoeboid, round paired, multiple or band form when observed under the microscope.

2.9 Recording of blood parasites

If a parasite is seen on a slide it is focused properly all-round the field of view to note the density of the parasites and was recorded. (Appendix V).

2.10 Removal of ticks from hosts

Ticks were removed by forcible detachment as described byJames-Rugu and Iwuala (1992, 1995, 1997, 1998 and 2001). This was done after the owners restrained the animals. Care was taken so that the appendages of the ticks are not damaged. The collected ticks were then transported to the laboratory in a labeled bottle.

2.11 Preparation of thick smears from tick vectors

The species of ticks collected from the animals were sorted out. The tick was placed on a clean grease-free microscope slide, a pair of forceps was used to hold the tick firmly on the slide, another pair of forceps was used to puncture the tick, after which the tick is returned to a labeled bottle containing 70% alcohol. The blood on the slide is then spread using the forceps. The method on thick smear preparation mentioned above was adopted. Care was also taken during puncturing not to damage the tick.

2.12 Fixation and identification of blood parasites from tick vectors

The method described above was adopted. This was followed by identification of the parasites from blood.

CHAPTER THREE
RESULTS AND ANALYSIS

During the study a total of 1,146 ticks were collected from 300 dogs and were identified at the species level. Three species of ticks all belonging to the family Ixodidae were identified (Table 2). *Rhipicephalus sanguineus* was the most abundant species encountered. This was followed by *Boophilus decoloratus* and *Haemaphysalis leachii.*

Ticks were prevalent throughout the monitoring period, regardless of the amount rainfall. However, the overall average tick load per dog was higher during the rainy months than during the dry months. The same was true for the average load of the three dominant tick species identified in the study. The rainy period also coincide with lower maximum and higher minimum monthly environmental temperature.

Out of the 300 dogs examined. 101 were found to be infected with *Babesia canis* giving 33.70% infection rate (Table 2) there was no record of other species of babesial parasite.

Table 3 showed prevalence of *Babesia canis* in relation to different breeds of dogs. Out of the three breeds examined mixed breed were the most parasitized with 37.20% infection rate. Next were the Alsatian breeds with 36.30% and finally the Mongrels with 18.00% infection rates. A statistical test (Chi-square) showed that at 0.05 level of confidence (p>0.05), there was no significant difference in the infection rates of the different breeds of dogs (Appendix I)

Table 4 indicated prevalence of *Babesia canis* in relation to the sexes of dogs. *Babesia canis* infection was higher in female dogs 37.70% than the male dogs 29.90%. It was found however, using a statistical test (chi-square) at 0.05 level of confidence (P>0.05), that there was no significant difference in the infection rates between the sexes of dogs (Appendix II).

Table 5 showed the prevalence of *Babesia canis* in relation to the age of dogs. It was found that young dogs had the highest infection rate with 35.00%. The infection rate does not differ appreciably from that of Adolescent having 33.80% and the Adult 32.60% which is the lowest infection rate recorded.

A total of 561 ticks were collected from different breeds of dogs. The species of ticks observed and identified were all drown from the family Ixodidae. Three species of ticks were noted namely; *Rhipipcephalus sanguineus, Haemaphisalis leachii* and *Boophilus decoloratus* (table 6). Results obtained showed that 4.96% of the ticks were confined to the mongrels where as 1.30% and 1.21% occurred on Mixed breed and Alsatian breed respectively (Table 7). The mean occurrence of different species of ticks on different breeds of dogs is also presented on Table 7.

Table 8. showed the infection rate of *Babesia canis* from the different tick species with *Rhipicephalus sanguineus* having the highest parasite load of 27.63% followed by *Boophilus decoloratus* with 22.82% while *Haemaphisalis leachii* have the lowest parasite load of 17.65%.

Table 9. Presented a record of *Babesia canis* from dog ticks in different breeds of the dogs. Mongrels have the highest infection of *Babesia canis* with

26.70%, it was followed by the mixed breed with 22.82%, and Alsatian had the lowest infection of 18.54%.

Table 10: Population of tick species encountered on dogs during   the study

Tick species	Total No. encountered	% occurrence
Rhipicephalus sanguineus	465	40.58
Haemaphysalis leachii	297	25.92
Boophilus decoloratus	384	33.50
Total	1146	100.00

Table 2: Infestation and mean occurrence of ixodid ticks on different breeds of dogs

Type of Breed	Number examined	No. with ticks (%)	Total No. of A (Mean)	Total No. of B (Mean)	Total No. of C (Mean)	TotalNo.of Ticks (Mean)
Mongrel	50	38(76.00)	205(4.10)	120(2.40)	148(2.96)	473(9.46)
Mixed	137	65(4745)	155(1.13)	95(0.69)	130(0.95)	380(2.77)
Alsatian	113	52(46.02)	105(0.93)	82(0.72)	106(0.93)	293(2.59)
Total	300	155(51.67)	465(1.55)	297(0.99)	384(1.28)	1,146(3.82)

Key A = *Rhipicephalus sanguineus*
B = *Haemaphysalis leachii*
C = *Boophilus decoloratus*

Table 3: Infestation of ixodid ticks in relation to the breeds and sex of dogs

Breed of dogs	Sex of dogs	Number examined	No. with ticks (%)	Total No. of A (Mean)	Total No. of B (Mean)	Total No. of C (Mean)	Total No. of Ticks (Mean)
Mongrel	Male	26	18(69.23)	90(5.00)	60(3.30)	72(4.00)	222(12.33)
	Female	24	20(83.33)	115(5.75)	60(3.00)	76(3.80)	251(12.55)
	Total	50	38(76.00)	205(5.40)	120(3.20)	148(4.00)	473(12.45)
Mixed	Male	70	30(42.86)	55(1.80)	45(1.50)	65(2.20)	165(5.50)
	Female	67	35(52.24)	100(2.86)	50(1.40)	65(2.00)	215(6.14)
	Total	137	65(47.45)	155(2.40)	95(1.50)	130(2.00)	380(5.85)
Alsatian	Male	58	22(37.93)	40(1.80)	40(1.80)	50(2.20)	130(6.00)
	Female	55	30(54.54)	65(2.20)	42(1.40)	56(1.87)	163(5.40)
	Total	113	52(46.02)	105(2.02)	82(1.60)	106(2.04)	293(5.63)
Grand Total		300	155(51.67)	465(3.00)	297(1.90)	384(2.50)	1,146(7.40)

Key A = *Rhipicephalus sanguineus*
  B = *Haemaphysalis leachii*
  C = *Boophilus decoloratus*

Table 4: Infestation of ixodid ticks in relation to the breeds and sex of dogs

Breed of dogs	Sex of dogs	Number examined	No. with ticks (%)	Total No. of A (Mean)	Total No. of B (Mean)	Total No. of C (Mean)	Total No. of Ticks (Mean)
Mongrel	Young	10	7(70.00)	42(6.00)	11(1.60)	16(2.30)	69(10.00)
	Adolescent	25	18(72.00)	63(3.50)	15(0.83)	13(0.72)	91(5.05)
	Adult	15	13(86.70)	58(4.50)	12(0.92)	18(1.40)	88(8.00)
	Total	**50**	**38(76.00)**	**163(4.30)**	**38(1.00)**	**47(1.24)**	**248(6.53)**
Mixed	Young	30	15(50.00)	22(1.50)	10(0.66)	9(0.60)	41(3.00)
	Adolescent	75	30(40.00)	40(1.33)	13(0.43)	18(0.60)	71(2.40)
	Adult	32	20(62.5)	32(1.60)	12(0.60)	16(0.80)	60(3.00)
	Total	**137**	**65(48.15)**	**94(1.45)**	**35(0.54)**	**43(0.66)**	**172(2.65)**
Alsatian	Young	20	12(60.00)	20(1.70)	10(0.83)	12(1.00)	42(2.70)
	Adolescent	48	25(52.10)	35(1.40)	12(0.48)	15(0.60)	62(2.40)
	Adult	45	15(33.33)	22(1.50)	4(0.30)	11(0.70)	37(2.50)
	Total	**113**	**52(46.02)**	**77(1.50)**	**26(0.50)**	**38(0.73)**	**141(2.70)**
Grand Total		**300**	**155(51.67)**	**465(3.00)**	**297(1.90)**	**384(2.50)**	**1,146(7.40)**

Key A = *Rhipicephalus sanguineus*
    B = *Haemaphysalis leachii*
    C = *Boophilus decoloratus*

Table 5: Prevalence of ticks in relation the breeds and mode of life of dogs

Breed of dogs	Mode of life	Number examined	No. with ticks (%)	Total No. of A (Mean)	Total No. of B (Mean)	Total No. of C (Mean)	Total No. of Ticks (Mean)
Mongrel	Penned	15	8(53.33)	50(6.25)	30(3.75)	48(6.00)	128(16.00)
	Free ranging	35	30(85.70)	155(5.20)	90(3.00)	100(3.33)	345(11.50)
	Total	**50**	**38(76.00)**	**205(5.40)**	**120(3.16)**	**148(4.00)**	**473(12.45)**
Mixed	Penned	100	30(30.00)	55(1.83)	15(0.50)	30(1.00)	100(3.33)
	Free ranging	37	35(94.60)	100(2.86)	80(2.30)	100(2.86)	280(8.00)
	Total	**137**	**65(47.45)**	**155(2.40)**	**95(1.50)**	**130(2.00)**	**380(5.85)**
Alsatian	Penned	80	20(25.00)	35(1.75)	12(0.60)	6(0.30)	53(2.65)
	Free ranging	33	32(97.00)	70(2.20)	70(2.20)	100(3.13)	240(7.50)
	Total	**113**	**52(46.02)**	**105(2.02)**	**82(1.60)**	**106(2.04)**	**293(5.63)**
Grand Total		**300**	**155(51.67)**	**465(3.00)**	**297(1.90)**	**384(2.50)**	**1,146(7.40)**

Key A = *Rhipicephalus sanguineus*
    B = *Haemaphysalis leachii*
    C = *Boophilus decoloratus*

Table 4: *Babesia canis* infection from different species of ixodid ticks

Tick species	Number of ticks examined	Number of ticks with *B.canis* (%)
Rhipicephalus sanguineus	465	443(95.27)
Haemaphysalis leachii	297	282(95.00)
Boophilus decoloratus	384	367(95.60)
Total	1146	1092(95.30)

Key: *B.canis* = *Babesia canis*

Table 5. *Babesia canis* recorded from dog ticks

Types of breed & No. examined	Species of Ticks	Number examined	Number with *B. canis* (%)
Mongrel	*R. sanguineus*	205	195(95.12)
	H. leachii	120	111(92.50)
(50)	*B. decoloratus*	148	144(97.30)
	Total	**473**	**450(95.74)**
Mixed	*R. sanguineus*	155	150(96.80)
	H. leachii	95	92(96.80)
(137)	*B. decoloratus*	130	128(98.50)
	Total	**380**	**370(97.40)**
Alsatian	*R. sanguineus*	105	98(93.33)
	H. leachii	82	79(96.34)
(113)	*B. decoloratus*	106	95(89.62)
	Total	**293**	**272(92.83)**
Total		**1146**	**1092(95.30)**

Key: *R. sanguineus* = *Rhipicephalus sanguineus*

   H. leachii   = *Haemaphysalis leachii*

   B. decoloratus = *Boophilus decoloratus*

   B. canis = *Babesia canis*

Table 6: Single species of ticks on different breeds of dogs

Species of ticks	Mongrel (38)			Mixed (65)			Alsatian (52)			Total
	No. infested	% infested	Total ticks (Mean)	No. infested	% infested	Total ticks (Mean)	No. infested	% infested	Total (Mean)	Total
R. sanguineus	6	15.80	100(2.60)	15	23.10	95(1.46)	10	19.20	55(1.06)	250
H. leachii	4	10.50	48(1.30)	5	7.70	52(0.80)	5	9.62	25(0.50)	125
B. decoloratus	5	13.20	80(2.10)	10	15.40	67(1.03)	10	19.20	50(0.96)	197
TOTAL	15	39.50	228(6.00)	30	46.20	214(3.30)	25	48.10	130(2.50)	572

Key: R. sanguineus = Rhipicephalu sanguineus

H. leachii = Haemaphysalis leachi

B. decoloratus = Boophilus decoloratus

Table 7: Multiple species of ticks on different breeds of dogs.

Species of ticks	Mongrel (38)			Mixed (65)			Alsatian (52)			Total
	No. infested	% infested	Total ticks (Mean)	No. infested	% infested	Total ticks (Mean)	No. infested	% infested	Total ticks (Mean)	
R. sanguineus & H. leachii	10	26.30	96(2.53)	10	15.40	90(1.40)	8	15.40	45(0.90)	464
R. sanguineus & B. decoloratus	10	26.30	100(2.60)	17	26.20	96(1.50)	13	25.00	55(1.06)	317
R. sanguineus, H. leachii & B. decoloratus	3	8.00	33(0.86)	8	12.30	34(0.50)	6	11.50	25(0.50)	365
Total	23	60.50	229(6.00)	35	53.85	220(3.40)	27	51.90	125(2.40)	574

Key: R. sanguineus = Rhipicephalu sanguineus

H. leachii = Haemaphysalis leachi

B. decoloratus = Boophilus decoloratus

Table 9: Overall occurrence of tick species sampled during the study

Month (2008)	Species of ticks	Number encountered	%occurrence
JANUARY	*R. sanguineus*	20	39.22
	H. leachii	13	25.50
	B. decoloratus	18	35.30
	Total	**51**	**100**
FEBRUARY	*R. sanguineus*	20	37.04
	H. leachii	16	29.63
	B. decoloratus	18	33.33
	Total	**54**	**100**
MARCH	*R. sanguineus*	28	36.80
	H. leachii	26	34.20
	B. decoloratus	22	29.00
	Total	**76**	**100**
APRIL	*R. sanguineus*	35	37.60
	H. leachii	28	30.10
	B. decoloratus	30	32.30
	Total	**93**	**100**
MAY	*R. sanguineus*	35	35.00
	H. leachii	28	28.00
	B. decoloratus	37	37.00
	Total	**100**	**100**

Key: *R. sanguineus* = *Rhipicephalu sanguineus*

H. leachii = *Haemaphysalis leachi*

B. decoloratus = *Boophilus decoloratus*

Table 9: Overall occurrence of tick species sampled during the study

Month (2008)	Species of ticks	Number encountered	% occurrence
JUNE	R. sanguineus	40	36.70
	H. leachii	30	27.50
	B. decoloratus	39	35.80
	Total	**109**	**100**
JULY	R. sanguineus	50	40.00
	H. leachii	31	24.80
	B. decoloratus	44	35.20
	Total	**125**	**100**
AUGUST	R. sanguineus	63	43.20
	H. leachii	34	23.30
	B. decoloratus	49	33.50
	Total	**146**	**100**
SEPTEMBER	R. sanguineus	77	46.40
	H. leachii	37	22.30
	B. decoloratus	52	31.30
	Total	**166**	**100**
OCTOBER	R. sanguineus	77	43.00
	H. leachii	37	24.00
	B. decoloratus	52	33.00
	Total	**226**	**100**
	Grand total	**1,146**	

Key: R. sanguineus = Rhipicephalu sanguineus

H. leachii = Haemaphysalis leachi

B. decoloratus = Boophilus decoloratus

Table 8a: Monthly Infestation of tick on different breeds of dogs

Breeds (Number)	Months	Temp °C	RH	Rainfall (mm)	Species of ticks	Number infested	% Infested
Mongrels	JANUARY	11	17	-	*R. sanguineus*	10	43.50
					H. leachii	7	30.40
(38)					*B. decoloratus*	6	26.10
						23	100
	FEBRUARY	10	20	-	*R. sanguineus*	7	39.00
					H. leachii	5	28.00
					B. decoloratus	6	26.10
						18	100
	MARCH	19	25	-	*R. sanguineus*	10	37.00
					H. leachii	10	37.00
					B. decoloratus	7	26.10
						27	100
	APRIL	19	30	103.7	*R. sanguineus*	15	38.00
					H. leachii	14	36.00
					B. decoloratus	10	26.00
						39	100
	MAY	17	35	150	*R. sanguineus*	13	41.00
					H. leachii	12	37.50
					B. decoloratus	7	22.00
						32	100

JUNE	16	60	250	R. sanguineus	18	43.00
				H. leachii	15	36.00
				B. decoloratus	9	21.00
					42	100
JULY	16	62	337	R. sanguineus	12	33.00
				H. leachii	10	28.00
				B. decoloratus	14	39.00
					36	100
AUGUST	16	70	345	R. sanguineus	25	46.00
				H. leachii	10	18.50
				B. decoloratus	19	35.20
					54	100
SEPTEMBER	16	75	280	R. sanguineus	28	51.00
				H. leachii	12	22.00
				B. decoloratus	15	27.30
					55	100
OCTOBER	16	65	200	R. sanguineus	37	45.00
				H. leachii	20	24.40
				B. decoloratus	25	30.50
					82	

Table 8b: Monthly Infestation of ticks on different breeds of dogs

Breeds (Number)	Months	Temp °C	RH	Rainfal (mm)	Species of Ticks	Number infested	% Infeste
Mixed	JANUARY	11	17	-	R. sanguineus	6	35.30
					H. leachii	5	29.40
(65)					B. decoloratus	6	35.30
						17	100
	FEBRUARY	10	20	-	R. sanguineus	5	29.40
					H. leachii	6	35.30
					B. decoloratus	6	35.30
						17	100
	MARCH	19	25	-	R. sanguineus	12	40.00
					H. leachii	10	33.00
					B. decoloratus	8	27.10
						30	100
	APRIL	19	30	103.7	R. sanguineus	10	39.40
					H. leachii	14	41.20
					B. decoloratus	10	29.40
						34	100
	MAY	17	35	150	R. sanguineus	12	29.00
					H. leachii	14	34.00
					B. decoloratus	15	37.00
						41	100

JUNE	16	60	250	*R. sanguineus*	12	32.00
				H. leachii	10	27.00
				B. decoloratus	10	40.00
				32	100	
JULY	16	62	337	*R. sanguineus*	25	49.00
				H. leachii	11	21.00
				B. decoloratus	15	29.00
				51	100	
AUGUST	16	70	345	*R. sanguineus*	25	46.00
				H. leachii	14	26.00
				B. decoloratus	15	28.00
				54	100	
SEPTEMBER	16	75	280	*R. sanguineus*	32	46.00
				H. leachii	13	18.00
				B. decoloratus	25	36.00
				70	100	
OCTOBER	16	65	200	*R. sanguineus*	30	40.50
				H. leachii	19	25.70
				B. decoloratus	25	34.50
				74	100	

Table 8c: Monthly Infestation of ticks on different breeds of dogs

Breeds (Number)	Months	Temp °C	RH	Rainfall (mm)	Species of Ticks	Number infested	% Infeste
Alsatian	JAN	11	17	-	R. sanguineus	4	36.40
					H. leachii	1	9.10
(52)					B. decoloratus	6	54.50
						11	100
	FEB	10	20	-	R. sanguineus	8	42.00
					H. leachii	5	26.00
					B. decoloratus	6	32.00
						19	100
	MAR	19	25	-	R. sanguineus	6	32.00
					H. leachii	6	32.00
					B. decoloratus	7	36.00
						19	100
	APR	19	30	103.7	R. sanguineus	10	50.00
					H. leachii	0	0.00
					B. decoloratus	10	50.00
						20	100
	MAY	17	35	150	R. sanguineus	10	37.00
					H. leachii	2	7.40
					B. decoloratus	15	55.60
						27	100

JUN	16	60	250	*R. sanguineus*	10	33.00
				H. leachii	5	17.00
				B. decoloratus	15	50.00
				30		100
JUL	16	62	337	*R. sanguineus*	13	34.20
				H. leachii	10	26.30
				B. decoloratus	15	39.00
				38		
AUG	16	70	345	*R. sanguineus*	13	34.20
				H. leachii	10	26.30
				B. decoloratus	15	39.50
				38		100
SEP	16	75	280	*R. sanguineus*	17	41.40
				H. leachii	12	29.30
				B. decoloratus	12	29.30
				41		100
OCT	16	65	200	*R. sanguineus*	30	43.00
				H. leachii	15	21.00
				B. decoloratus	25	36.00
				70		100

Table 10: Over all prevalence of *Babesia canis* in dogs.

Babesial infection	Number examined	%
Infected	101	33.70
Uninfected	199	66.30
Total	**300**	**100.00**

Table 11:   Prevalence of *Babesia canis* in different breeds of dogs

Breed	Number examined	Number infected	% infected
Mongrel	50	9	18.00
Mixed	137	51	37.20
Alsatian	113	41	36.30
Total	300	101	33.70

Table 12: Prevalence of *Babesia canis* in different sexes of dogs.

Sexes	Number examined	Number infected	% infected
Male	154	46	29.90
Female	146	55	37.70
Total	300	101	33.70

Table 13: **Prevalence of *Babesia canis* in different age group.**

Age group	Number examined	Number infected	% infected
Young	60	21	35.00
Adolescent	148	50	33.80
Adult	92	30	32.60
Total	300	101	33.70

CHAPTER FOUR
DISCUSSION AND CONCLUSION

The predominant species of ticks observed and identified were *Rhipicephalus sanguineus, Haemaphisalis leachii* and *Boophilus decoloratus*. The ticks encountered and recorded in this study have been documented as ticks of dogs in Plateau State (Tanwia, 1982, 1989, James-Rugu and Iwuala 1992, 1994, 1998 and James-Rugu 2001). The result indicates 27.63% 0f *Rhipicephalus sanguineus* were infected with *Babesia canis* while 17.65% *Haemaphisalis leachii* and 22.82% of *Boophilus decoloratus* were also noted to be infected with *Babesia canis*.

The presence of these blood parasites in these ticks is a well-known fact (Hoogstraal, 1956) and it therefore confirms that these ticks are vectors of these blood parasites as was also reported by (James-Rugu, 2001).

The mongrels (Local) showed higher infestation rate and mean occurrence of Ixodid ticks of 248 (4.96%) compared to the Mixed 172(1.30) and Alsatian 141(1.24%). This could be as a result of much attention given to the exotic breeds than the mongrels as a result of income accrued through the exotic breed which is far more than that of Mongrels. Medical attention is given to the exotic breeds than the Mongrels. Most mongrels are free ranging which expose them to have frequent contact with ticks; this is different in the case of the exotic breeds.

There is an interesting issue about this investigation. Despite the fact that most of the dogs sampled showed no signs of illness yet, they harbored some babesial parasite as shown in table 2. This means that most of the dogs were in a state of premunity with persistent low grade parasitemia, a phenomenon which is common in babesial infection and has been known to last for years. Premunition does not offer complete protection of the dogs from the disease. Hence, a greater challenge as any stress factor such as poor nutrition would trigger the disease. This is a risk to the dog owner as this disease of dogs is of zoonotic importance (Sobczyk *et al.*, 2005).

The disease being zoonotic meaning it can affect man as well, has not been reported, however, literature have shown that in other parts of the world Brasil, Germany, Taiwan, New York and Korea the disease has been reported in man. The percentage of the infection in dogs appears to be increasing in Plateau with the increasing man and dog relationship it therefore means the disease may be found in men probably in small percentage (Hunfeld *et al.*, 2002 and Maia *et al.*, 2007).

This study revealed a high prevalence of *Babesia canis* in dogs despite the fact that none of the dogs showed clinical signs of the infection. Smyth (2005) had earlier on reported that most of the tick-borne organisms caused inapparent infections but could induce clinical disease when conditions are adverse to the host. Losos (1986) also states that an important characteristic of babesial infection is the long duration, which indicates that organisms can survive in the host in spite of the immunological responses. This ability may partly be due to the antigenic variation which occurs with each succeeding parasitemia during chronic phases of infection and has been known to last for years. This work is in conformity with the findings of Eseyin (1996) who recorded a high prevalence of *Babesia canis* in domestic dogs. It compares favourably with that of Ogunkoya *et al.*, (1986) who, worked on blood parasites of dogs and noted that *Babesia canis* was the highest haemoparasitic infection recorded from dogs. Kumbin (1988) had reported a high infection rate of *Babesia canis* in her study on haemoparasites of domestic dogs. It also compares favorably with the work of James – Rugu (2001) who reported high infection rate of *Babesia canis*. Kamani *et al.*, (2008) also reports a high infection rate of *Babesia canis* in their studies on Parasitic causes of anaemia in dogs in Vom and environment. This higher incidence could be attributed to some climate or environmental factors favouring the distribution of ticks which are vectors of this parasite. Such factors include lower mean monthly temperature, high annual rainfall and plenty vegetation cover with greater moisture content serving as habitat for these ticks.

Within the breeds, mixed breed had the higher infection than the other breeds. Alsatian breeds closely followed and the mongrels had the lowest infection rate of. It then means that the mongrels may be more resistant to infection than the exotic breeds. These differences conform to the findings of Odewumi *et al.*, (1979) who recorded a lower infection rate in mongrels compared with the higher infection in exotic breeds in Enugu and Nsukka zones of the eastern states of Nigeria. Similarly Eseyin (1996) worked on the prevalence of Babesial parasites of domestic dogs in Jos Area of Plateau state and reported a low prevalence rate of *Babesia canis* in mongrels and higher in exotic breeds. These differences may be explained by the fact that the mongrels may have had prior and longer exposure to babesial parasites and therefore developed a degree of tolerance as evidenced by the significantly lower occurrence of clinical disease amongst them. This may also be associated with other factors like hair condition and hair colour which, aid the attachment of the intermediate host.

Female dogs had higher infection rate than the male dogs. Tanwia (1989) had earlier on reported a similar finding that the female dogs carried greater loads of ticks than the male dogs. It is probable that these ticks are potential carriers of the parasite. This result may be due to sex differences and possibly the stress encountered during pregnancy in some cases. It can also be explained when a consideration is made on the activities of the female dogs. Female dogs appear to be more active and mobile than the male dogs and as a result there is a corresponding increase the possibility of the female being more exposed to ticks and thus more parasitized than the male dogs.

Incidence of *Babesia canis* was slightly higher in the young dogs than in either the adolescent or adult. The infection rate was slightly higher in young dogs than in adult. Young dogs are therefore more susceptible to babesial infections than adult dogs. The fact that young dogs are more agile and their high rate of activity which make them come in frequent contact with tick infestation may be an explanation to the infection rate mentioned above. Adult dogs on the other hand are like confined to their environment.

The high incidence of this infection and parasitic disease common to man and his animals together with the increased recognition of their grave impact on the health and well-being of people, on reduced production, and on retarded development explains the interest on this topic.

RECOMMENDATION

The application of preventive measures against transmission of infection from animals to man is an important issues to be looked at. Infected dog can serve as a reservoir of the infection to man. It is therefore necessary to take measures both preventive and curative against the infection. Dogs should be taken to the clinic for regular routine check-up. This study was carried out to exert a favourable effluence toward bringing about a change in the attitude of pet owners not taking their pets for routine check-up.

This disease may be prevented through avoidance of ticks and regular routine bathing of the dog with chemical for the removal of ticks. A strict programme of tick control as is done for Rabies in dogs remains the most reliable preventive measures against the disease. There should be mass education of people on the public health hazards of the disease. Veterinarians and Zoologist who have the knowledge of the disease should therefore extend their services to the urban communities as well as the rural areas in advising pet owners on the hygiene management and nutrition of pets. There should be regular blood examination and prompt effective and efficient treatment of dogs with appropriate dose in endemic areas. Blood donors should be screened for *Babesia canis* before transfusion is done. Affected animals should undergo quarantine that is, isolation for treatment and after treatment. Since most of the dogs are free ranging there should be a little restriction to avoid contact with ticks.

Control may be established by systematic elimination of ticks in domestic animals. By artificially establishing a state of premunition with blood inoculation, it is possible to protect animals being taken to enzootic areas. Careful monitoring of inoculated animals is always necessary because of difficulties in establishing standardized dozes.

A vaccine against a specific, avirulent canine babesiosis strain (Piriodo®) is available in France; however, cross-immunity between the different strains of *Babesia* appears not to occur. It has been shown recently that protective immunity to a virulent strain is possible thus, making vaccines from different parasite strains feasible.

Premunity has been recognized as important in controlling clinical signs of the virulent form of the disease in endemic areas, therefore complete eradication of parasites from infected animals may not be advantageous in these areas; and thus the use of drugs to sterilize the infection may be undesirable. The role that premunity plays in areas with less virulent strains is not known.

REFERENCES

Adeyanbu, B.J, and Abdulahi, U.S. (1986). Clinical Syndromes due to babesiosis, Ehrlichiosios and Hepatozoonosis in Dogs. 2nd National conference on Haemoparasitic Disease and their Vectors. Abstract No. 34.

Ameh, J.E (2007). A study on the Ectoparasites of Dogs in some part of Jos Plateau State. *BSc. Thesis, Department of Zoology, University of Jos, Jos Nigeria.*

Anon (1986). Home care of the sick or injured pet. Published by Hills pet products, Ltd U.K. pp.1-80.

Anon (1986) Preventive Health care for Dog. Published by Hill's pet products Ltd. UK pp1-100.

Banetti, G., Kenny, M.J., Tasker, S., Anug, Y. Shkap, V., Levy, A and
  Bansal, S.R., Gautam, O.P., and Baneslee, D.P. (1985). prevalence of *Babesia canis* and *Hepatozoan canis* infection in Dogs of Hirsor (HARYANA) and Delhi and attempts to isolate *Babesia* from Human beings. *Indian Veterinary Journal,* 62:748 – 751.

Birkenheuer, A.J., Neel, J., Ruslander, D., Levy, M.G. and Breitschwerdt, E.B. (2004). Detection and molecular Characterization of a novel large Babesia species in a dog. *Veterinary Parasitoloigy*, 124:151-160.

Breitschwerdt, E.B., Malone, J.B., Macwilliams, P., Levy, M.G., Qualls Jr, C.W. and Prudich, M. J. (1983). Babesia in the Grey hound. *Journal of the American Veterinary Medical Association,* 182 (9): 978-979.

Camacho, A.T., Pallos, E., Gestol, J.J., Giuitian, F.J. and Olmeda, A.S. (2001). *Babesia canis* infection in splenectomized Dog. *Bulletin of Sociology and Pathology of Exotic,* 94(1): 17-19.

Chang, S.H., Park, J.H., Kwak, J.E., Joo, M., Kim, H., Chi, J.G., Hong S.T. and Chai, J.Y. (2006). A case of histologically diagnosed tick infestation on the scalp of a Korean child. *Korean Journal of Parasitology*, 44 (2): 157-161.

Chung, Y.M. (2006). Studies on ticks and tick-borne parasites of some breeds of Dogs in Jos South Local Government Area of Plateau State. *B.Sc. Thesis, Department of Zoology, university of Jos, Nigeria.*

Comazzi, S., Paltrinieri, S., Mangedi, M.T. and Agnes, F. (1999). Diagnosis of *Canine babesiosis* by percoll gradient separation of parasitized erythrocytes. *Journal of Veterinary Diagnosis and Investigation*, 11:102 – 104.

De-Castro, J.J. (1997). Sustainable ticks and tick-borne diseases control in livestock improvement in developing countries. *Veterinary Parasitology*, 71:77-97.

Farwell; G.E., Legrand, E.K., and Cobb, C.C. (1982). Clinical observations on *Babesia gibsoni* and *Babesia canis* infections in Dogs. *Jouranl of the American Veterinary Medical Association*, 180 (5): 507-511.

Foldvari, G., Marialigeti, M., Solymosi, N., Lukacs, Z., Mojoros, G., Kosa J.P., and Farkas R. (2007). Hard Ticks Infesting Dogs in Hungary and their infection with Babesia and Borrelia species *Parasitology Resource*, 101:525-534.

Foster, R. and Smith, M. (2008). Ticks: Life cycle, Anotonny and Disease Transmissions. http://www.petEducation.com/dog.

Georg, B.D.J. (2003). Comparative study of Haemocyte populations in *Babesia* sp. Infected and unaffected *Boophilus decoloratus* (koch) Ticks. *Nigerian Journal of Entomology*, 20:49-55.

Hall, H.T.B. (1977). Disease and parasites of Livestock in the Tropics Longman Group Limited, Pp 1-278.

Harwood, R.F. and James, M.T. (1969). *Entomology in Human and Animal Health*. Macmillan Company, Canada, 371pp.

Hoogstraal, H. (1956). *Ticks of the Sudan*. U.S. Naval Medical Research Unit (3) Cairo, Egypt,1110pp.

Hunfeld, K.P. and Brade, V. (2004). Zoonotic Babesia: possibly emerging pathogens to be considered for tick-infested humans in central Europe. *International Journal of Medical Microbiology*, 293 (39): 93-103.

Hunfeld, K.P., Lambert, A., Kampen, H., Albert, S., Epe, C., Brade, V. and Jenter, A.M. (2002). Seroprevalence of Babesia infections in Humans exposed to ticks in Midwestern Germany. *Journal of Clinical Microbiology*, 40 (7): 2431-2436.

Iwuala, M.O.E. and Okpala, I. (1978b). Studies on the Ectoparasitic fauna of Nigerian Livestock II: Seasonal infestation rates. *Bulletin of Animal Health and Production in Africa*, XXVI (4): 361 -358.

45

Iwuala, M.O.E., and Okpala, I. (1978a). Studies on the Ectoparasitic fauna of Nigerian Livestock I: Types and distribution patterns. *Bulletin of Animal Health and Production in Africa,* XXVI (4): 339-350.

James – Rugu, N.N. (2002). Studies on ticks and tick borne parasites of Dogs in Jos, Plateau state. *Zuma Journal of Pure and Applied Sciences,* 4 (2): 29-35.

James – Rugu, N.N. (2003). Experimental studies on *Boophilus decoloratus* infestation and tick-borne zoonotic infections of dogs in Jos Plateau State, Nigeria. *International Journal, Environmental Health and Human Development,* 4(2): 56-62.

James - Rugu, N.N., (2001). A study of the haemoparasites of Dogs, pigs, and cattle in Plateau State, *Nigerian Journal of Science and Technology,* 7: 20-27.

Kamani, J., Sannusi, A., Dogo, G.I., Egwu, O.K., TankoJ.T., Kemza, S. and Onovoh, E. (2008). Parasitic causes of anaemia in dogs in Vom and environment. *Nigerian Journal Parasitology, Abstract. No.5.*

Konnai, S., Saito, Y., Nisenikado, H., Yamoda, S., Imamura, S., Mori, A., Ito, T., Onuma, M. and Oshashi K. (2008). Establishment of a Laboratory colony of Taiga tick *Ixodes persulcatus* for tick-borne pathogen transmissions studies. *Japanese Journal of Veterinary Research,* 55 (2-3): 85-92.

Kumbin, M.Y. (1988). Studies on the blood parasites of domestic dogs- Canis familiaris in Jos Area of Plateau state. *B.Sc. Thesis Department of Zoology, University of Jos, Nigeria.*

Losos, G.J. (1986). *Infectious Tropical Disease of domestic animals.* Published in Association with the International Development Research Centre. Canada, 450pp.

Maia, M.G., Costa, R.T., Haddad, J.P.A., Passos, L.M.F., and Ribeiro, M.F.B. (2007). Epidemiological aspects of canine babesiosis in the semiarid area of the state of Minas Gerais, Brazil. *Preventive Veterinary Medicine* 79:155-162.

Mery, L. (2006). History and background of dog. http://historyofdog.com

Mohammed, A.N. and Agbede, R.I.S. (1980). Control of ectoparasites on ruminants in Nigeria. In proceedings of national seminar on " The current problem facing the Leather Industry in Nigeria. LERIN, Zaria, September, 24-26 1980.

46

Obori, G. (2001). A complete Guide to Dog care Ganob and Associates   pp1-30.

Ogunkoya, A.B., Ogunkoya, Y.O., Sarara, D.I. and Ogunsusi, R.A (1986). Prevalence of Blood parasites and effects on the Haemogram of Dogs. *2nd Nationla Conference on Haemoparasitic Disease and their Vectors. Abstract No. 15.*

Oliver, J. (1994). The complete Book of Dog care published by Parragon Book services Ltd., UK.

Onuoha, O.E. (2006). Prevalence of ticks and tick-borne parasites of some breeds of dogs in Jos – South Local Government Area of Plateau state. *B.Sc. Thesis, Department of Zoology, University of Jos, Nigeria.*

Oyamada, M., Davoust, B., Boni, M., Deruere, J. Bucheton B., Hammad, A., Itamoto, K., Okuda, M. and Inokuma, H. (2005). Detection of *Babesia canis rossi, B canis vogeli* and *Hepatozoan canis* in Dogs in a village of Eastern Sudan by using a screening PCR and sequencing methodologies. *Clinical Diagnostic Laboratory Immunology*, 12 (11): 1343-1346.

Oyerinde, J.P.O. (1999). *Essentials of Tropical Medical Parasitology.* University of Lagos press. Akoka Lagos, Nigeria, 321pp.

Penzhom, B.L. and Chaparro, F. (1994). Prevalence of *Babesia Cynicti* infection in three populations of yellow mongoose (cynistis penicillata) In the transvaal, South Africa. *Journal of wild life Diseases*, 30 (4): 557-559.

Shih, C.M., Liu, L-P., Chung, W-C., Ong, S.J. and Wang, C-C (1997). Humana Babesiosis in Taiwan: Asymptomatic infection with a Babesia microti – like organism in a Taiwanese woman. *Journal of Clinical Microbiology*, 35(2): 450 -454.

Show, S.E. (2004). Infection with a proposed new subspecies of *Babesia canis, Babesia canis* sub-species Presentii in domestic cats. *Journal of Clinical Microbiology*, 42 (1): 99-105.

Siegmund, O.D. (1998). the Merck Veterinary Manual 8th Edition, merck and co. Inc. White house station N.J. USA, Pp 19-1209.

Smyth, J.D. (2005). *Animal parasitological* Cambridge low-price editions, pp.1-549.

Sobczyk, A.S., Kotomski, G. Gorski, P. and Wedychowio, H. (2005). Usefulness of Touch down PCR Assay for the Diagnosis of a typical cases of *Babesia canis canis* infections in Dogs. *Bulletin of Veterinary Institute Pulowy,* 49: 407-410.

Soulsby, J.I. (1982). Helminths, arthropods and protozoa of domesticated animals, 7th Edition William and Wilkins company, Baltimore. USA 121-155.

Tanwia, N.N.A. (1989). Aspects of the biology and ecology of ticks of livestock in some parts of Plateau State. *Ph.D. thesis Department of Zoology, University of Jos, Nigeria.*

Telmadarraiy, Z., Babarimi, A. and Vatandoost, H. (2004). A survey on Founa of Ticks in West Azerbaijan province, Iran. *Iranian Journal of Public Health,* 33(4): 65-69.

Vredevoe, L. (2006). Background Information on the Biology of Ticks. http://entomology.ucdavis.edu/faculty/rbkimsey/ticbiol.html.

White, D.J., Talaruco, J., Chang, H.G., Burkhead, G.S., Hermberger, T. and More, D.L. (1998). Human Babesiosis in New York state. *Archive of International Medicine,* 158: 2149 -2154.

WHO (1991). Basic Laboratory methods on Medical Parasitology. World Health. Organization, Geneva.

Zeleke, M. and Beleke, T. (2004). Species of Ticks on Camels and their seasonal population dynamics in Eastern Ethiopia. *Tropical Animals Health and production,* 36:225-231.

APPENDICES
Appendix I
Chi-square (χ^2) Analysis to test the discrepancies in infection rates of dogs

Breeds	Number Examined		Number Infected		Total
	Obs.	Exptd	Obs	Exptd	
Mongrel	50	44.14	9	14.90	59
Mixed	137	140.65	51	47.40	188
Alsatian	113	115.20	41	38.80	154
Total	300		101		401

$\chi^2 = 3.6489$.

Table value = 5.99

Tabulated value at $0.05 = 5.99$ is greater than the calculated value (3.649) so we conclude that there is no significant difference in the infection rates between the different breeds of dogs.

Appendix II
Chi-square (χ^2) Analysis to test for discrepancies in infection rates of different sexes of dogs.

Sexes	Number Examined		Number Infected		Total
	Obs.	Exptd	Obs	Exptd	
Male	154	149.63	46	50.37	200
Female	146	150.37	55	50.63	201
Total	300		101		401

$\chi^2 = 1.0109$

Table value = 3.84
Tabulated value at 0.05 is 3.84 and is greater than the calculated value (1.0109) we conclude that there is no significant differences in the infection rates between the different sexes of dogs.

Appendix III
Chi-square (χ^2) Analysis to test for discrepancies in the infection rates of different Age-groups of dogs.

Age - group	Number Examined		Number Infected		Total
	Obs.	Exptd	Obs	Exptd	
Young	60	60.60	21	20.40	81
Adolescent	148	148.13	50	49.90	198
Adult	92	91.30	30	30.73	122
Total	300		101		401

$\chi^2 = 0.0465$

Table value = 5.99
Tabulated value at 0.05 is 5.99 and is greater than the calculated value (0.047) we conclude that there is no significant difference in the infection rates between the different Age-groups of dogs.

Department of Zoology,
Faculty of Natural Sciences,
University of Jos.

SKB/NNJ-R

Appendix IV: Questionnaire for the sample collection.

S/No.	Breed of dog	Age	Sex	Mode of life	Remark

Key:

Sex   M = Male
       F = Female

Department of Zoology,
Faculty of Natural Sciences,
University of Jos.

SKB/NNJ-R

Appendix V: Examination of blood sample

S/No.	Breed of dog	Age	Sex	Species of ticks		
				A	B	C

Key:

Sex	M = Male	A= *Rhipicephalus sanguineus*
	F = Female	B= *Haemaphysalis leachii*
		C= *Boophilus decoloratus*

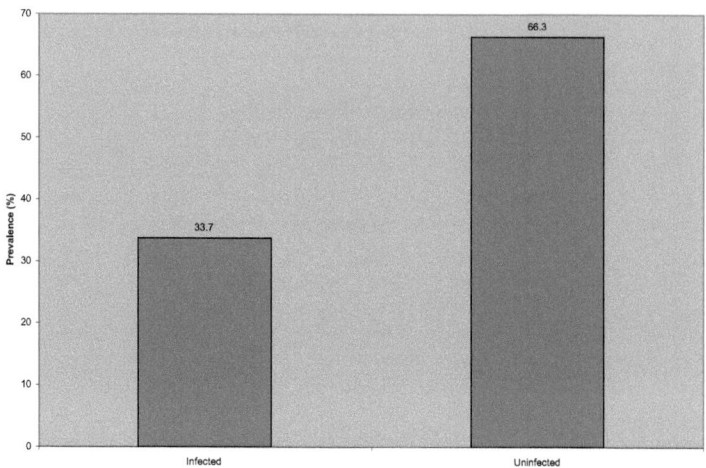

Appedix VI: Overall prevalence of Babesia canis in dogs.

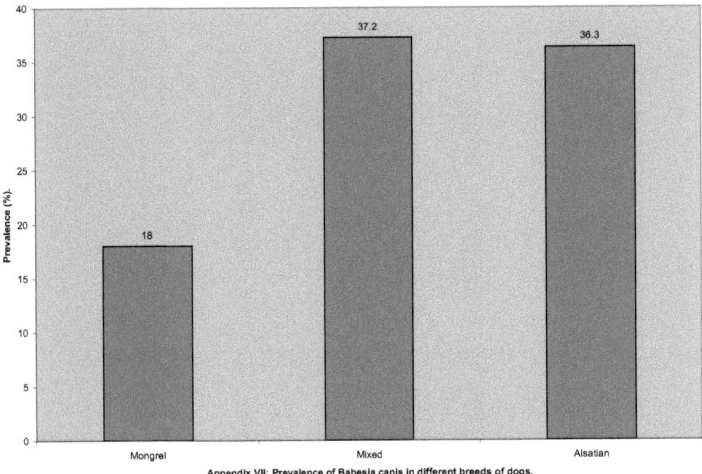

Appendix VII: Prevalence of Babesia canis in different breeds of dogs.

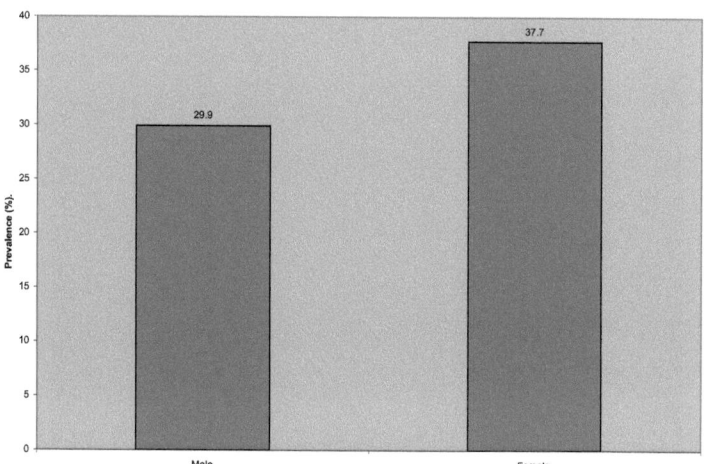

Appendix VIII: Prevalence of Babesia canis in different sexes of dogs.

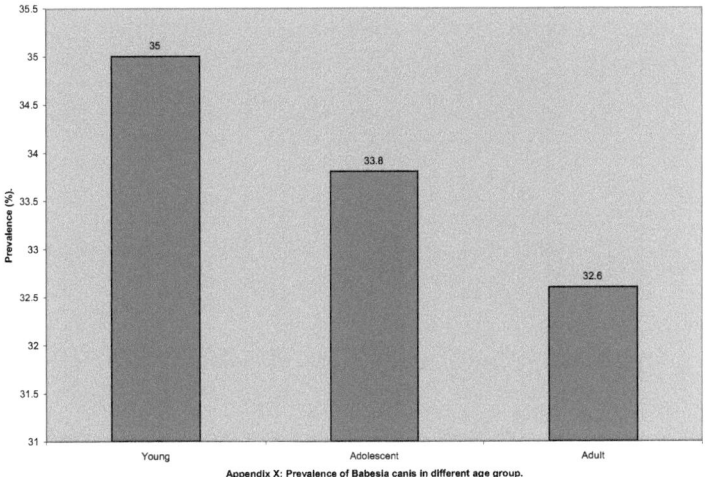

Appendix X: Prevalence of Babesia canis in different age group.

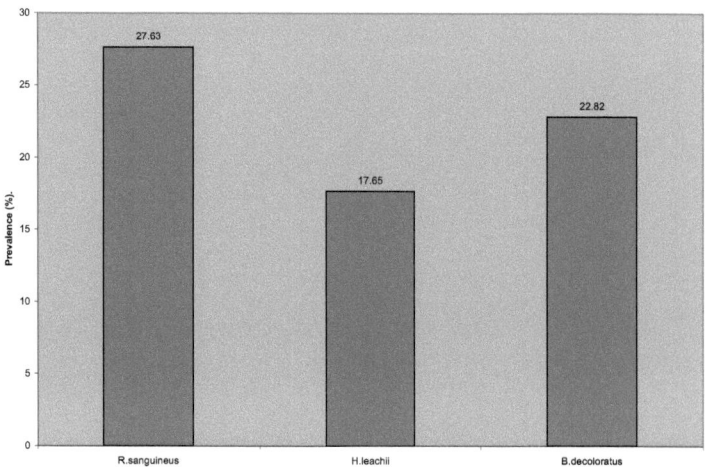

Appendix IX: The relative abundance of tick species during the study period.